Acid Rain, Acid Snow

John Slade

ACID RAIN, ACID SNOW

ISBN 1-893617-08-4

Library of Congress Control Number
00-090935

FIRST EDITION

All photographs by the author.

Printed on recycled, acid-free paper
at
Canterbury Press
Rome, New York 13440

WOODGATE INTERNATIONAL
P.O. Box 190
Woodgate, New York 13494

This book is dedicated to the peepers.
May we open our windows to hear them in the spring, forever.

Acknowledgements

A warm and hearty thank you to Marilyn Thomas, volunteer at the Woodgate Library, for her generous contribution to the research for this book. Marilyn, you are an Adirondack angel.

Many excellent people have helped with the research for this book. I offer my deep gratitude to:

Kjetil Tørseth, Arne Semb, and Jan Schaug at the Norwegian Institute for Air Research, in Kjeller, Norway. Tusen takk!

Peter Bauer of the Residents' Committee to Protect the Adirondacks, in North Creek, New York.

David Greenwood of the Adirondack Council, in Albany, New York.

The Adirondack Museum, in Blue Mountain Lake, New York, which organized an excellent conference on acid rain in August, 1999.

Walter Kretser of the Department of Environmental Conservation, in Ray Brook, New York.

Edward Bennett of the Department of Environmental Conservation, in Albany, New York.

Gerrit Snyder of the Attorney General's Office, in Albany, New York.

Lynn Igoe of Mohawk Valley Community College, in Utica, New York.

David Goldstein and John Mimikakis of New York Representative Sherwood Boehlert's office, in Washington, D.C.

Matt Little and D. Gray Maxwell of New York Senator Daniel Patrick Moynihan's office, in Washington, D.C.

Jim Kellogg of the Water Quality Lab, Richard Poirot of the Air Quality Lab, and Eric Titrude of the Attorney General's Office, in Vermont.

Kathleen Fallon Lambert of the Hubbard Brook Research Foundation, in Hanover, New Hampshire.

Marcia Argust of the National Parks Conservation Association, in Washington, D.C.

Jonathan Banks of Clear the Air, in Washington, D.C.

Christine Real de Azua of The American Wind Energy Association, in Washington, D.C.

Adam Serchuk of the Renewable Energy Policy Project, in Washington, D.C.

Christer Ågren, editor of "Acid News," the excellent newsletter from The Swedish NGO Secretariat on Acid Rain, in Göteborg, Sweden.

Contents

Other books by John Slade

DANCING WITH SAMUEL
A JOURNEY OUT OF DARKNESS
CHILDREN OF THE SUN
HERBERT'S MOUNTAIN
THE NEW ST. PETERSBURG
A DREAM SEEDED IN THE EARTH
GOOD MORNING, DADDY!

Available from
Woodgate International
P.O. Box 190
Woodgate, New York 13494
315-392-4508

The Friend Whom I Love is Dying

When I was a year old, my grandparents built a camp on the shore of a small lake in the Adirondack Mountains of northern New York State. As the summers passed, I learned to dog paddle in that lake, while my mother cupped her hand beneath my tummy; I learned to swim, back and forth between my mother and father; and then I learned to dive from the dock into deep water, with a real dive and not a belly flop.

Wearing my orange life jacket, I learned how to row an old barge of a rowboat.

Best of all, when I was ten years old, (in 1957,) I learned how to look through a skin diver's mask, and how to breathe through a snorkel, which I had bought with my allowance at Woolworth's. Kicking slowly with a pair of flippers around a cluster of lilypads, I explored the lake's busy underwater world, . . . until my skinny body shivered, my lips turned purple, and my teeth chattered so much that I could hardly tell my family back on the dock what I had seen.

I stared down at a submerged stump with writhing roots: a stump that looked like a huge octopus. I circled nervously around a beaver house, and became twice as nervous when I discovered its door, an underwater opening in that heap of sticks. (Would a beaver come surging out, its chisel-sharp teeth ready to chomp my femur?) And I treasured an old cracked teacup which I found on the bottom, filled with black Adirondack muck; no good for tea, it was perfect for holding wild raspberries when I went picking with my grandmother.

We all know that lilypads are green on top, but how many know that they are shiny red underneath? Their tangled stems spiral down to the black bottom. Cruising along the edge of a lilypad patch, I watched a school of perch drifting and darting among the slender stems, some of them two inches long, (aquarium size,) and some up to ten inches long, (the size my grandfather and I liked to catch with a worm on a hook).

Schools of shiners also lived in the lilypad forest: a glint of silver as they darted in the sunshine, and then they were gone.

Pumpkinseeds, however, would hover in front of me, broadside, the red spot on their gill cover bright in the sunshine. If, with a slight kick of my flippers, I advanced toward a pumpkinseed, it would advance toward *me*, hover head-on a few inches from the glass of my mask, and defiantly defend its territory.

Yellow-brown bullheads would skitter across the bottom, then settle with their whiskers spread on the muck. If I reached down toward them, they would shoot off with a thrashing tail, leaving a swirl of black debris

After meandering around a patch of lilypads, my second favorite place to snorkel was along the edge of a floating bog. A bog is a floating sponge of sphagnum moss a foot or more thick, with several feet of open water beneath it. A bog is thus a floating island, perhaps five feet across, perhaps fifty, anchored to the bottom of the lake by a tangle of long roots.

Growing on top of the bog, in the sunshine, are a great variety of plants that are neither terrestrial nor aquatic; rather, they take root in the damp, slightly acidic sphagnum sponge. Pitcher plants and sundews catch insects and digest them. A species of rose, and a species of orchid, both require the special conditions on the bog; they would die if transplanted to the earthen shoreline of the lake. Red maples sprout and grow into saplings on a bog. Tamaracks, also known as larches, may grow over ten feet tall, even though their roots have a sodden foothold.

The tamaracks lose their needles in winter. In the spring, their dead-looking skeletons sprout a whole new set of needles, a lovely pale green in May, dark green by July, and a radiant yellow in September. The tamaracks take different shapes as they grow in the wind out on the lake. Scattered around the broad, irregular bog, they look, especially when viewed from a canoe in the morning mist, like a troupe of dancers whirling in green gowns through the mystical choreography of some ancient rite.

But all that is on top of the bog. Beneath it, the variable layer of open water sandwiched between the bog and bottom is in the bog's shadow, laced with a thick tangle of roots, just the sort of place where a snorkeler might find the pale, puckered, well-preserved body of someone who drowned a century ago.

I would grab hold of the outer edge of the bog, (my hand gripping the sunlit moss,) then I would lower my head and snorkel beneath the bottom of the bog and reach in as far as I could among the roots, to where the water was cold and dim. I was wary of getting tangled in those murky twisted roots, and of stirring up the black muck with my flippers as I struggled to free myself, . . . until, growing desperate for air, I would dig my way upward through the bog itself, but my snorkel is caught in the roots, a tree is growing just above me, and I can't . . . glug, glug . . .

Never mind, I always survived. And often I would spot, while peering under the bog, a colony of freshwater sponge growing on those roots. Most people think that sponges grow only in the ocean, but one species grows in Adirondack lakes. Beneath a bog, the sponge grows as a tube around the roots; at the edge of a bog, the sponge may grow around the twigs of a submerged branch. Under the bog, the sponge is pale gray. But at the edge of the bog, where sunshine penetrates the water, algae grows symbiotically inside the sponge, and the gnarly witch's fingers are green.

While snorkeling along the edge, I occasionally encountered a large bass hovering in the shadow beneath the bog, facing outward toward the sunlit amber water, waiting for a frog to hop into the lake. As I approached with a twitch of my flippers, the bass would swing toward me, its pectoral fins fanning with agitation, ready to charge an aggressor into its territory. I would drop the toes of my flippers, (thus putting on the brakes,) and hover with my heart thumping at arm's length from the sleek, dark, fierce-eyed lunker.

If I dared to wiggle a finger at it, the bass would dart into the depths and vanish. I was left with a swirl of muck from the bottom of the bog.

Often, while snorkeling in the warm sunlit water along the edge of a bog, I was delighted to discover a school of baby bass. In the early summer, these hatchlings were like thin, delicate slivers of glass, less than an inch long, swimming in schools of hundreds. Later in the summer, they became darker and bulkier, with a distinct orange stripe on their tail; the schools were now smaller, with fifty to a hundred miniature bass. By late summer, the fingerlings, up to two inches long, swam as individual fish along the rocky shoreline and beneath the docks.

Every summer, I would set up an Adirondack aquarium in my grandparents' camp. The gravel and plants came right from the lake. Once the under-gravel filter had cleared the water, I would stock the aquarium with young bass, pumpkinseeds, perch, and bullheads. I caught these one- and two-inch fry with a minnow trap placed in the shallows overnight.

While snorkeling, I would pick up red ramshorn snails for the aquarium. From a canoe, I would sweep a net back and forth, catching a multitude of tiny creatures, including daphnia and water boatmen and bright orange water mites. Everybody found a new home in the aquarium, and everybody seemed content, except for my grandmother, who complained that the hum of the aeration pump made a loud buzzing in her hearing aide.

A few days after setting up the aquarium, I would survey its glass walls with a magnifying glass. Delicate white hydras, the freshwater cousins of corals and sea anemones, trailed their quarter-inch tentacles in the water, seeking to capture one-celled protozoans. Snails laid their eggs on the glass, where I could watch the tiny embryos develop. I could also watch the rhythmically munching mouths of adult snails as they cruised (or oozed) across the glass, rasping with their tongues a lunch of algae.

I often added a crayfish to the community, a caddisfly larva, and a few shiny black whirligigs. The crayfish would wave its claws like the neighborhood bully, catching bits of chopped earthworm that the fish no longer bothered with after they had feasted. The caddisfly larva would crawl across the gravel, then disappear beneath a stone. The whirligigs zoomed around the surface, banging into the glass. (And thus I soon returned them to the lake.)

Water boatmen, insects with a pair of legs like oars, dove down and clung to the plants, carrying with them a shiny silver bubble.

Dozens of daphnia, each the size of the head of a pin, swam in their jerky, random manner; I could just barely see their dark eyes, as big as the point of a pin.

I spent hours peering into that aquarium, making new discoveries every day. My brother and sister would join me to watch the bass and pumpkinseeds snap at dead flies that we dropped onto the surface. Visiting neighbors would stare with fascination at tadpoles beginning to sprout their hind legs, at arrow-headed planaria (tiny flatworms) slithering along the glass, and at the green fingers of sponge.

My grandfather liked that glass box of lamplit plants in his living room, especially at night when we turned off the other lights. He had built the camp, he had made much of the furniture, and he approved of the aquarium as a worthy addition.

My grandmother, however, preferred to stay in her kitchen, baking a huckleberry pie and crabbing about that darn buzzing pump.

I have snorkeled in my beloved Adirondack lake every summer for forty-three years. Gradually, I saw fewer fish. Fewer snails. I could turn over dozens of rocks and sunken logs, and find no crayfish.

In 1959, while snorkeling for an hour, I would see hundreds of fish, especially in the lilypad jungles. Snails and snail eggs were on nearly every sunken stick. The swarms of daphnia were so thick, like clouds of dancing fairy motes, it seemed as if the small fish would never run out of an abundant dinner.

Forty years later, in 1999, I could snorkel for an hour and see five small perch among the lilypad stems, a few pumpkinseeds hovering above their bowl-shaped nests in the sandy bottom, and perhaps a single bass lurking beneath the edge of a bog. But I no longer saw bullheads, or snails, both of which are bottom feeders. Nor did I see water boatmen with their shiny silver bubble, or even a few sparse daphnia. The plants in the lake looked good, but most of the animal life had vanished.

Though I occasionally encountered a school of bass hatchlings along the edge of a bog in the spring, by midsummer the schools were greatly reduced, with less than fifty inch-long fish. By the end of the summer, I saw very few orange-tailed fingerlings anywhere in the lake. Further, I rarely saw a midsized bass, (six to ten inches long,) which may indicate that the larger adults are still spawning, but the few fingerlings which manage to live through the summer do not survive their first winter.

My neighbors in their camps along the shoreline, many of them staunch in their determination that the lake is "just fine," tell me that their grandchildren caught a whole passel of pumpkinseeds that summer, that a pair of big bullheads often appear in the shallows near their dock, that the pike (which somebody put into the lake) or the snapping turtles must be eating everything. The fishermen on the lake contribute to a program of stocking with five-inch bass; as long as the fishing is good, the lake must be fine.

And of course, the sun still sparkles on the surface. The water-ski boats and jet-skis roar back and forth as if life just couldn't get any better.

When I set up an aquarium during the summer of 1999, I covered the bottom with gravel and some waterlogged sticks, then rooted or floated several kinds of plants. I let the water settle for several days, then I explored with the magnifying glass: I found no hydras, no daphnia, no arrow-headed flatworms with their tiny eyespots.

While snorkeling, I couldn't find a single snail to put into the aquarium. I left the minnow trap out overnight, with bits of bread in it; morning after morning it was empty. I tried setting it in the lilypads; and hanging it by a string from the edge of a bog; and setting it in the shallows behind an island. Empty. Empty. Empty.

As a snorkeler in the lake for over forty years, I have explored every bay and beaver house, have slowly circumnavigated nearly every bog and island, and have cruised along extensive stretches of the shoreline. During the past decade, I have surveyed the sandy bottom where a stream pours into the lake, and I no longer see the bowl-shaped nests made by bass. I have flippered through aquatic gardens of pickerelweed, and I no longer see even a single tadpole. I have held my breath and dived down into the chilly depths beside a long granite wall, where I explored a tangle of huge waterlogged trees that toppled years ago from the cliff above; I no longer see a dozen large bass hovering in the shadows beneath the logs, but maybe one.

I have come to love this lake as a friend which has always welcomed, summer after summer, the lovelorn teenager, the confused college student, the frazzled graduate student, the tired teacher, the disillusioned taxpayer. I love the lake because I learned to swim in it, learned to paddle a canoe on it, learned to cast a Tiny Crazy-Crawler so that it splashed precisely beside a half-sunken log, where surely a monster bass must be lurking.

On a gorgeous September day, I took my grandmother, wrapped in a blanket, out in the canoe so that she could see the geese one last time. When my mother was failing with Alzheimer's, she still loved to go out in the canoe after dinner, paddling in the bow with her featherweight paddle, and she still jumped with startled delight when a beaver suddenly slapped its tail. Those were some of her last smiles, looking back at me from the bow and whispering, "KaPOOSH! "

I love this Adirondack lake because it so generously welcomes my friends who come to visit it, from California, from Manhattan, from Iceland, from Russia. For the foreigners especially, this beautiful little lake tucked deep in the forest of the Adirondack Mountains becomes a major part of their experience in America.

I don't tell my Russian friends that anything is wrong. I don't tell them, as they step gingerly from the canoe onto a bog for the first time, that there are almost no frogs left. I don't tell them that the kingfishers, with their blue crest and rattling call, rarely dive into the lake anymore. I don't tell them that perch and bass in many Adirondack lakes are now tainted with mercury. (Rather, I ask them to toss whatever they catch back into the lake, "so we can catch him again.") I don't tell them that the trout are dying because aluminum destroys their gills.

I let my Russian guests breathe the pine-scented air, and listen to the cheerful call of chickadees in the hemlocks. From my porch, the Russians watch a blazing orange sun as it descends behind the feather-duster white pines across the lake. After dinner, we paddle the canoe up the channel into the wilderness end of the lake, with the hope of hearing a beaver slap its tail. We sleep at night with rain drumming on the roof, and for breakfast we eat my lumpy lopsided pancakes covered with genuine New York maple syrup.

I do not tell my Russian friends that most sugar maple seedlings now die before they become saplings. I don't say that where thousands of young trees used to carpet the floor of a sugar maple forest, only a few scattered individuals now survive beneath the canopy of elders.

I let my happy Russian friends take back to the polluted cities of Murmansk and St. Petersburg their memories of "paradise in America." I let them take back with them their hope that one day we will work, all of us together, to make the entire world as healthy as it is in the Adirondacks.

To Lave: To Wash Clean

Water laves its way through the world,
Touching all life:
It courses up the vascular pipeline to the top of the tallest sequoia.
Laces through the leaves of the rustling maple,
And drifts as perfumed mist from the petals of a pink apple blossom.
Water oozes as a protoplasmic soup,
Cleansing and feeding and helping to propel the peripatetic amoeba.
And water races as a red river through a tiny tubular canal,
Cleansing and feeding and helping to propel the peripatetic thoughts
Wandering among the cells of your brain.
From capillary to continental river,
From thunderstorm to heartbeat,
Water laves its way through the world.

800-Foot Smokestacks

A veteran whom I know came home from Vietnam in 1960. One of the first things he did here in America was to seek a place of peace and beauty: he headed deep into the Adirondack Mountains to his favorite trout stream. But though he fished for several days, he did not catch a single trout.

Senator Daniel Patrick Moynihan of New York, in his October 1998 testimony before Congress on acid rain, stated "As far back as the 1960's, fishermen in the Adirondacks began to complain about more than 'the big one that got away.' Fish, once abundant in the pristine, remote Adirondack lakes, were not just getting harder to catch. They were gone."

Senator Moynihan continued, "When I entered the Senate in 1977, there was much we needed to learn about acid rain. So I introduced the first Federal legislation to address our 'knowledge deficit' about acid rain: the Acid Precipitation Act of 1979. My bill was enacted into law as Title VII of the Energy Security Act, which Congress passed in June 1980 (Public Law 96-264). Title VII established the National Acid Precipitation Assessment Program (NAPAP), an interagency program charged with assessing the causes and damages of acid deposition, and reporting its findings to Congress."

Between 1980 and 1990, 570 million dollars worth of research was conducted under the NAPAP program. By October of 1989, more than 1,700 research papers had been published in technical journals on the subject of acid precipitation. After weapons research and space research, the study of acid rain "had become the Federal government's biggest scientific undertaking in history."

The winds blow across America from west to east. When the coal-burning power plants in the Midwest were required by the Clean Air Act of 1970 to reduce their air pollution, which for decades had been blowing across the eastern United States, the plants met regional standards by building taller smokestacks, up to 800 feet in height. Now, much less of their smoke soiled the towns and cornfields of Ohio; instead, most of the smoke blew eastward on the upper winds, so that greater amounts of pollution blew further east than ever before.

By mixing with the moisture in the air, sulfur and nitrogen in the coal smoke transformed into sulfuric acid and nitric acid, H_2SO_4 and HNO_3. This acidic water vapor rose to a higher altitude when it reached the western slope of the Adirondack Mountains, and it grew colder as it rose higher. Condensing, the vapor fell to earth as either rain or snow; or it wrapped as a cloud around the mountain peaks. The sulfuric and nitric acids effected nearly everything they touched, including the granite bedrock.

The information gathered during the 1980's was used to take legislation a step further. Senator Moynihan was vital to that next step: "I was an original co-sponsor of the Clean Air Act Amendments of 1990," which were "a radical departure from the traditional 'command and control' approach to environmental regulation common at the time. This program was the first national . . . market-based approach to pollution control. It has been tremendously successful."

Electrical power plants which released less than their quota of sulfur through their smokestacks were able to sell that amount of unreleased sulfur to other power plants, which could then exceed their limits by that amount. Power plants were thus economically encouraged to lower their sulfur emissions, so they could earn an extra profit. This system worked extremely well in most regards, for sulfur emissions fell, by 1995, to a level that was nearly 5 million tons of sulfur below the 1980 level, which was 39 percent ahead of the program's goal.

Equally positive, the cost to the power plants proved to be far less than they had estimated in 1990. Though electric utilities had complained that it would cost them well over a billion dollars to comply with the 1990 regulations, actual costs, as determined by the Massachusetts Institute of Technology, totaled $726 million during the first five years. In 1990, utilities protested that it would cost them well over a thousand dollars for every ton of sulfur they removed from coal or its smoke. However, the actual cost has proven to be well under a hundred dollars per ton, reaching a low of $68 per ton in March 1996.

Though substantial progress has been made with sulfur pollution, acid rain continues for a number of reasons. First, the opportunity to purchase a greater allowance of sulfur emissions enabled a large number of companies to do just that. Rather than clean up their smoke, they bought a higher pollution limit, and thus they continue, quite legally, to let the wind carry away their filth. The 1990 program has resulted in a 17 percent reduction of sulfur emissions nationwide, but this reduction is far too small to solve the problem.

Second, the Clean Air Act Amendments of 1990 did not limit nitrogen emissions with a national cap. There is no trading of allowances to encourage a reduction of nitrogen emissions. Therefore, though the amount of sulfuric acid has been reduced in the rain and snow falling on the Adirondacks, the amount of nitric acid has increased.

Third, the Amendments of 1990 said nothing about mercury emissions. This trace component of coal was also drifting eastward in the smoke, and in combination with acidic water, it took on a form that was far deadlier. The mercury worked its way up the Adirondack food chain until it was stored in the fatty tissue of many fish, especially smallmouth bass and perch.

In August 1998, NAPAP presented a report to Congress on its acid rain research: the report is a 144-page compilation of data, charts and text based on eighteen years of intensive study. The report stated that although sulfur emissions had been reduced, this level of reduction was not enough to enable Adirondack lakes to recover from the harm they had suffered, and continued to suffer, from acid precipitation. In fact, at the current level of sulfur and nitrogen emissions, a growing number of lakes in the Adirondacks would acidify, and die.

Consequently, Senator Moynihan introduced Senate Bill 172, the Acid Deposition and Ozone Control Act, which would further reduce sulfur emissions by 50 percent, and require nitrogen emission reductions of 70 percent, from power plants. This bill recommends the same "cap-and-trade" program for nitrogen that has already proven successful for sulfur. It also requires the Environmental Protection Agency to regulate the amount of mercury in power plant emissions.

Sulfur and nitrogen come not only from the coal and oil used to fuel electrical power plants, but from gasoline as well. The cars, trucks, planes and trains of America produce almost half of the NOx (nitrogen oxide) pollution carried by the wind. The exhaust from your tailpipe and mine ends up, in part, as sulfuric acid and nitric acid in our rain.

Senator Moynihan has therefore introduced Senate Bill 171, the Clean Gasoline Act, which would require vehicles across the country to meet California sulfur standards of 30 parts per million, rather than the present national average of 330 parts per million.

This bill will also reduce nitrogen in tailpipe exhaust.

To summarize, we are moving in the right direction, but we have not gone far enough. Unless we move further in reducing sulfur and nitrogen pollution from our power plants, and from our millions of vehicles on the road, then our lakes and our forests, our rivers and streams, will become less and less able to support a rich community of life.

And when the canary in the coal mine stops singing, the miners are next.

Acid rain is not merely an Adirondack problem, but a national problem. From Maine to Florida, states downwind from the Mid-western power plants have suffered from decades of acid precipitation. The Appalachian Mountains, where much of the coal was originally mined, have been devastated first by the bulldozers and then by the returning smoke. Camel's Hump in Vermont now bristles with dead trees. Dying forests and dead streams are now common in the Catskill Mountains of New York, the Great Smokey Mountains of Tennessee, and the Allegheny Mountains of West Virginia.

Further west, the Rocky Mountains of Colorado suffer from acid rain caused by pollution from Mexico. The San Bernardino and San Gabriel Mountains within the Los Angeles air basin suffer from acid precipitation.

And Canadians are quick to inform us that more than half of their acid rain comes from the United States, blowing northeast across the long border, whereas only 17 percent of our acid rain comes from Canada.

The problem, in fact, is global. Like our Adirondack fishermen, Norwegian fishermen discovered a striking reduction of fish in their streams and lakes during the 1960's. The first international conference on acid precipitation was held in Stockholm in 1972. Countries throughout Europe now pool their information from extensive research, and the European Union has issued a series of increasingly strict protocols on pollution limits. But, as in the United States, these limits have proven to be insufficient: the lakes in Scandinavia, and the forests of Poland and the Czech Republic, are still dying. The Germans have their own word for it: "Waldsterben," or "forest death."

Acid precipitation effects countries around the world, including Russia, India, and Japan across the Asian continent, Brazil and Venezuela in South America, and Nigeria in Africa. China, which relies heavily on coal, experiences rain in some regions that is nearly as acidic as vinegar.

We have slowed the damage, but we have not stopped it, and we certainly have not reached the point where acidified lakes and forests can start becoming healthy again. At present, over 400 lakes and ponds in the Adirondack Mountains are dead, and 27 percent of all Adirondack lakes and ponds have a year-round pH (acidic level) of less than 5.0, which means that fish, as well as many other creatures, cannot live in them. A full 75 percent of high elevation spruce trees in the Adirondacks have died, and a greater proportion of trees at lower elevations have been seriously weakened.

In Canada, roughly 14,000 lakes have been seriously acidified, and a much greater number are at the brink of dying. In Norway and Sweden, 16,000 lakes are filled with crystalline water, and no fish.

In the United States, we produce 57 percent of our electricity from coal. We pride ourselves in our advanced technology—we have walked on the moon, we have sent space probes beyond the edge of our solar system, we have the biggest and fastest computers in the world—and yet we are still filling our air with a stink that Charles Dickens would recognize.

Looking deeper into this dismal situation, we learn something that is never mentioned in the 144-page NAPAP report of 1998. According to a less known report, "Air Pollution in Our National Parks," written in 1999, "Coal-burning power plants that were either built or under construction at the time of the 1977 Clean Air Act Amendments were granted an exemption, or 'grandfathered,' from meeting future Clean Air Act emission standards. These grandfathered plants were expected to retire at the end of their 30-year projected lives. Yet they continue to operate today, releasing four to 10 times more pollution than modern plants. These grandfathered plants constitute 52 percent of our country's energy generation, yet they produce 97 percent of the acid rain and haze-causing sulfur dioxide, 85 percent of the ozone smog-causing nitrogen oxide, and 99 percent of the toxic mercury pollution from the utility sector."

These old power plants were supposed to clean up, or eventually quit. However, there are today "hundreds of grandfathered power companies across the country." And even though emissions in modern plants have dropped, "the amount of electricity generated from coal-fired power plants increased by almost 16 percent between 1992 and 1998. Power plant SO_2 emissions have increased every year from 1996 through 1998, rising more than 10 percent over 1995 levels. In 1998, power plants emitted 1.26 million more tons of SO_2 than they emitted in 1995." In other words, both the old plants and the modern plants are producing more electricity than ever before, and the result, despite the Clean Air Act and its Amendments, is more pollution.

The National Park report continues, "8 of the 10 power plants in Virginia are grandfathered coal-burning utilities." As a result, "95 percent of Virginia's SO_2 and NOx are produced from its grandfathered coal-burning power fleet." As a further result, "currently, only half of Virginia's native brook trout streams are fully capable of supporting fish populations."

As always, deadly mercury sneaks into the picture. "Coal-burning power plants are the largest industrial source of mercury . . . . Up to 50 percent of the mercury emitted from coal-burning power plants can travel up to 600 miles from the plant." Most stunning of all: "Today, electric power companies are not required to meet any mercury emissions standards, even though they are our country's largest contributors to this pollution."

After traveling on the winds, mercury falls with the rain or snow into our lakes, where it "is converted to organic mercury by bacteria and other processes. Organic mercury is easily taken up by plankton and small fish, thereby making its way into the food chain." Mercury does not merely give us a tummy ache. It is a "neurotoxin that effects children and women of childbearing age." When our kids eat the bass that they were so excited to catch, they risk damage to their nervous system.

Some of the grandfathered power plants have not quit, nor have they modernized their pollution controls, but instead have actually expanded their output of electricity, and thus their pollution, contrary to the Clean Air Act. The Environmental Protection Agency in Washington has therefore initiated a lawsuit against 32 coal-burning plants in the Midwest, claiming that they "spent hundreds of millions of dollars in modifying plants, increasing their life and increasing their pollution. They (have) done this without applying for permits, without giving public notice, and without installing pollution-control technology."

Several states in the East have begun their own lawsuits as well. On September 15, 1999, Attorney General Eliot Spitzer of New York announced that he would sue seventeen coal-fired plants in Ohio, Indiana, Virginia, West Virginia, and Kentucky. These court cases will take time, but at least the effort has begun to shut down the worst offenders.

On October 14, 1999, New York Governor George Pataki initiated a process to mandate that all New York State utilities cut emissions for sulfur dioxide by 50 percent and nitrogen oxide by 70 percent, thus requiring New York to meet the standards set by the pending Moynihan bill. Since New York State is responsible for 7 percent of the SO_2 and 20 percent of the NOx in its own acid rain, this legislation will both benefit New York's lakes, (as well as lakes and forests downwind,) and provide an example to other states, and to the nation.

On May 1, 2000, the New York State Legislature in Albany passed a bill which would discourage New York State power companies from selling their pollution credits to power companies in other states. Between 1995 and 1998, New York utilities sold about 700,000 credits to polluters in other states. Albany's new bill would require New York power companies to pay a fine equal to their income from the sale of credits to fourteen other specified states; the revenues from these fines would be used to develop nonpolluting energy sources, such as wind and solar power, in New York State. New York's electrical companies are, of course, contesting the new bill.

Ultimately, we need stronger legislation at the federal level, for only Congress in Washington can force power plants across the country to meet the standards necessary to bring health back to our forests and lakes. Senator Moynihan's bill in the Senate, S. 172, co-sponsored by Senator Schumer, and Sherwood Boehlert's bill in the House of Representatives, co-sponsored by John E. Sweeney, both place an emissions cap on sulfur and nitrogen, and require the EPA to regulate mercury, in all 48 contiguous states and in the District of Columbia. The goal of these two important bills is to balance coal emissions with what our natural world can tolerate. Until we reach and maintain that balance, we worsen a problem that will take decades, if not centuries, to fully correct.

Let us be clear that we have the technology to do the job, and that the cost of doing it has always proven to be far less than what the power companies claimed it would be. In fact, American firms which specialize in pollution control are presently earning a profit in Germany, where they enable German power plants to operate at one-fourth the level of pollution of American plants, while producing the same amount of electricity, at no greater cost.

Many other countries are developing comparable innovative technologies. They know that coal is cheap, but ultimately, wind power and clean lakes and thriving forests and healthy children are a whole lot cheaper.

Turn off the lights you're not using. Turn off those office machines when you're not using them.

Write a letter to your local power company, asking them how much coal they burn. And why. (See sample at end of book.)

Goodbye to My Brothers

One by one, life disappears:
Bass, sugar maples, loons; tadpoles, perch, people.
I hope we meet again.
I hope we reassemble with a bit more wisdom
On another planet.
What dya think?
One chance in a billion billion.

The Effects of an Acid Bath

When coal is burned in a factory or electrical power plant, the smoke emerging from the smokestack carries sulfur in combination with oxygen, and nitrogen in combination with oxygen. In the atmosphere, these two compounds combine with hydrogen to produce sulfuric acid, H_2SO_4, and nitric acid, HNO_3. Many other types of molecules are carried in the smoke, especially carbon dioxide, CO_2, but we shall focus on the two major acids which, while drifting in the wind, combine with droplets of water to form acidic raindrops, acidic crystals of snow, and acidic mist.

Acidic rainwater, and in the spring acidic snowmelt, flood the soil with a bath of dilute acid. The water may seep deeper into the earth until it reaches the water table, or it may flow downhill in brooks and streams and rivers. The acidic water gathers in ponds and lakes, and in the temporary springtime pools in a forest. And everywhere this acidic rainwater travels, it changes the chemistry of much that it touches.

In soil, the acid leaches calcium, an essential nutrient for plants, and washes it away. Over the years, acid rain can thus kill an entire forest by slow starvation, even though the sun is shining.

As we have seen, the acid also leaches aluminum from the soil and rocks, and the mobilized aluminum now wreaks havoc: it damages the fine root hairs of trees and thus prevents them from absorbing nutrients and water; it reduces the growth of cambium in trees, thus slowing the production of new cells and living wood; in ponds and lakes, aluminum increases the precipitation of tiny dark particles of organic matter called humics, thus clearing the water and enabling sunlight to penetrate deeper, causing a bloom of algae; aluminum poisons microscopic zooplankton, thus destroying major links in the food chain; and it gathers in the gills of fish, causing mucus, deterioration, and suffocation.

Acidic water in soil damages the bacteria and fungi that live in the earth. Part of their role in nature's complex web is to break down organic matter and so provide nutrients for the more advanced plants. If that role is curtailed by acid rain, once again the trees slowly starve.

Acidic water transforms mercury into deadly methyl mercury, CH_3Hg, which is absorbed into the fatty tissue of fish. This mercury may subsequently poison the loons, herons, ospreys, and otters that eat those fish.

Lovely white snow can be especially deadly, because unlike rain, a blanket of snow may contain an entire winter's worth of acid precipitation: five months of pollution in a frozen state, which suddenly melts into the earth. This jolt of acid, lasting two or three weeks, abruptly changes the chemistry of the soil, streams and lakes at that critical time when animal life, awakening from the long dark winter, begins the springtime ritual of laying eggs. Fish eggs and frog eggs and salamander eggs and mayfly eggs, and a hundred other types of eggs, are bathed throughout the spring in water from snowmelt. These delicate creatures and their eggs have evolved for millions of years; now they must suddenly contend with a lingering bath of acid, which may damage the eggs, damage the hatchlings, and damage the food supply. Those peepers which we used to hear by the thousands in the woods every spring, happy love-smitten frogs singing so loudly that we could hardly sleep, are now a few dozen lovelorn survivors trying to keep the species alive.

The clouds which wrap around mountain peaks may be more acidic than the rain. A forest growing on the slopes can be drenched by an acidic mist for days at a time. The needles on conifers absorb moisture from the air, and thus the acid as well. The waxy surface on leaves may be broken down. Photosynthesis is diminished, and again the tree starves.

Acid rain may not be the direct cause of a tree's death. Rather, it weakens the tree, so that insects, disease, the cold of winter, a summer drought, or fungi are able to overwhelm a tree's resistance, until it dies. The same is true with a broad range of plants and animals: weakened by their immersion in a bath of acid, and by slow starvation, they become victims to disease or severe weather.

A great number of animals require calcium in their shells. However, acidic water reduces their ability to metabolize calcium, and thus their shells may be soft or improperly formed. A soft shell provides little protection for crayfish, mayfly and dragonfly larvae, and a myriad of other creatures. (Black fly larvae, unfortunately, seem to thrive in acidic water.) The fish and birds that eat these invertebrates may likewise suffer from a loss of calcium. The shells of bird eggs may be thin, and the chicks may not receive enough calcium for their bones.

Many soils are capable of buffering the acids in snow and rain; the deeper and richer the soil, the more likely that it contains compounds which can neutralize the acids. But some soils, especially on the slopes of mountains, are thin, and they lay upon a bedrock of granite and quartz. Their buffering capacity is thus limited, and eventually used up. Then even the slightest bit of acid in the rain or snow results in an abrupt increase of acidity in the soil, in streams, in lakes, and ultimately in the broad watershed below the mountain. Even if we stopped burning coal and oil and gasoline today, vast areas of mountain soil around the world would require decades, perhaps centuries, before their chemical composition returned to normal.

Water, the basic ingredient of life, now carries with it, wherever it goes, foreign materials which originate in lumps of coal and pools of petroleum. The trout, the frog egg, and the protozoans fare no better in their acidified watershed than we would in a room filled with acrid smoke. We would dash for the door, or throw open a window; but the frog eggs have no other place to go.

If the trout had a voice, they might call the springtime snowmelt a "massacre." If the mayflies had a voice, they might call it "genocide." If the daphnia had a voice, they might demand that we people be hauled before some sort of world court.

For who are we to so blithely kill?

Maybe acid rain has to hit home before we pay any attention.

Water from Adirondack wells may leach copper and lead from the plumbing, thus washing them right into your drinking glass. The copper can kill bacteria in your septic tank, so that human waste is not broken down, but further contaminates the water.

Sulfur dioxide, as well as sulfate particles, which are both components of coal smoke, can damage healthy lungs. In particular, they can make breathing difficult for people who have asthma. Mephistopheles thrived on sulfur fumes, but they don't do us any good.

Acid rain eats away at buildings, monuments and sculptures made of limestone or marble. The acid dissolves the calcium carbonate in these stones, so that walls become pocked with small holes, square edges become rounded, and carved details become coarsened. In Washington, D.C., where the rain has an average acidity of pH 4.2, which is more than ten times as acid as pure rain, many of the historic buildings are becoming pocked, corroded, crusted, streaked and discolored. The same is true of buildings around the world. In Athens, Greece, the

Acropolis has weathered more during the past half-century than it did during the first twenty-four centuries after it was built.

Perhaps this is enough gloom and doom for one chapter.

Alas, we're not quite done yet.

Keep in mind that as forests die, their ability to remove carbon dioxide from the air decreases, and thus the blanket of earth-warming pollution increases. Acid rain directly increases the greenhouse effect.

Keep in mind that as forests die, a livelihood dies with them. Let's put an Adirondack logger and an Ohio coal shoveler together at the same conference table, and then let us pay attention to who speaks of sustainable resources, and who speaks of digging it and burning it until you run out of it.

And keep in mind that as forests die, as lakes die, as streams and ponds and springtime pools filled with peepers all die, something in our soul dies with them.

On a Starry Night

Look at all that flash and tatter out there:
Nuclear sparks hurtling through a vacuum laced with waves of energy.
And spinning as happy as could be
In the midst of that vast celestial desert
Is Earth, blessed with water, blessed with land,
Blessed with a gentle, steady dose of sunshine:
An oasis which the Creator chose as a cradle.

Matter ordered into the form of life
Seems to be only a tiny fraction of all the matter in the universe.
Most matter seems to be unliving,
Unless we believe that fiery gasses and chunks of frozen nitrogen
Are alive.
But only a small portion of living matter
Thinks,
Unless we believe that viruses and vegetables and ancient sequoias
Can think,
And only a small, small portion of thinking matter
Ponders its origin,
Unless we believe that apes and elephants and the singing whales
Can ponder.
That leaves
Us.
We see no clear bridge between star and starfish,
Nor between starfish and human child,
And so we ponder how we came to be alive.

There must have been some force that challenged
The randomness of existence,
And won.

Our Willing Thralldom to Oil

Coal is not the only vestige from the Dark Ages that still haunts our present and curses our future. It has a twin, equally ugly, named Oil.

Though most of the sulfuric acid and some of the nitric acid in acid rain comes from the burning of coal, the bulk of nitric oxides in our atmosphere come from the burning of oil, especially in the form of gasoline. The world's half-billion cars and trucks, with thousands more built every day, are all contributors to acid rain. So are the buses, railroads, ships, and airplanes that roar and rumble around our planet. Add to them the motor scooters, motorcycles, motorboats, jet-skis, go-karts, four-wheelers, snowmobiles, and all the rest of our motorized toys. Then add the lawnmowers, snow blowers, farm tractors, generators, industrial engines, and a multitude of military vehicles to the list. All of them are burning some form of oil.

We are pumping out an enormous amount of nitric oxide, NO, from our gasoline engines. In the atmosphere, nitric oxide combines with oxygen to form nitrogen dioxide, NO_2, which then transforms when mixed with water vapor into nitric acid, HNO_3. But nitrogen dioxide does not stop there. In the presence of ultraviolet radiation in sunshine, it can transform oxygen, O_2, into ozone, O_3. Further, nitrogen dioxide can combine with a broad range of manmade hydrocarbons to form chemicals called aldehydes. (The formaldehyde which you remember from biology class is one example.) We are poisoning ourselves every time we turn the ignition key.

"But," you say, "I thought my car had strict emission standards."

It probably does. It may even meet California emission standards, among the strictest in the world. Engines certainly run cleaner now than they did before the Clean Air Act of 1970. But, there are far more cars on the road now, in America and around the world. And, Americans are driving further than ever before.

In addition, Americans are purchasing loophole vehicles which do not meet the low-emission standards required for mid-sized sedans. Light-duty trucks, which include the popular Sport Utility Vehicle, made up 46% of new car sales, even though they use more gas per mile than ordinary cars, and produce three times as much smog-forming exhaust. The SUVs are heavier, their engines are more powerful, and their pollution emissions are comparably greater. Though the SUV is built to carry a gang of kids to soccer practice, (with all their gear,) or a group of neighbors in a car pool to the office, (with all of their briefcases and laptops,) in reality, the SUV carries, on the average, only 1.7 people per mile. Most people still drive to work alone.

Why do we drive so much, and so blithely? Because gasoline is cheap. When adjusted for inflation, the American price for gasoline was lower in 1997 than it has been during any year since 1920. American gasoline is much cheaper than petrol (the British word for gasoline) in Europe. Yes, the OPEC oil producers may force up the price now and then, but increased American prices still remain far less than increased European or Japanese prices. So we continue to cruise.

The surprising fact is that the automobile industry did not necessarily need gasoline at all. Nikolaus Otto, the German engineer who built a four-stroke internal combustion engine in 1876, used alcohol as his fuel. Tests in America in 1907, which compared ethanol and gasoline engines, showed that ethanol offered higher engine compression, and thus greater horsepower. Alexander Graham Bell and Thomas Edison both recommended ethanol as a fuel.

Henry Ford built his first automobile with an engine fueled by ethanol, which be called "farm alcohol" because it could be made from the same farm crops used by distilleries: grain, potatoes, or corn. He believed so much that "fuel from vegetation" would be the "fuel of the future" that his Model A, built in 1929, had an adjustable carburetor, able to mix air with either alcohol or gasoline.

Thus, we might well have powered our automobiles during the twentieth century with a fuel that is far cleaner than gasoline, and which could have been produced by American farmers. Our farmers would have prospered, our foreign trade deficit would have been much lower, and we would not have been dependent on a fuel that comes primarily from Arabian deserts.

Consider further: during the twentieth century, Americans made a great number of technological advances. Orville and Wilbur Wright first lifted on wings in 1903; but now, after several generations of improved airplanes, the space shuttle circles the earth, placing telescopes and satellites into precision orbits. We once had to crank the telephone and ask Central to connect us; but now, after generations of improved telephones, we can make a call from a cell phone while flying over the

Atlantic Ocean in a Concorde, and speak with our boss in New York and our customer in Paris at the same time. Though by 1900 we had moved beyond the abacus, we hadn't moved all that far; but now, after generations of calculators and computers, kids in kindergarten can surf the global Web. And though we once treated infectious diseases with cold compresses and hot tea, we now have a broad range of antibiotics. We perform eye surgery with a laser beam. We can even give a child a new heart.

Why then are we still burning gasoline in a four-stroke internal combustion engine, as we did a century ago? Is it really because oil is the one and only fuel, or has someone somewhere along the way impeded progress?

The answer to that complex question is beyond the scope of this short book. Nevertheless, because we Americans made so much of the hydrocarbon soup in the air, we bear on our shoulders a substantial responsibility for cleaning it up. We are less than 5 percent of the world's population, (we are roughly 4.6 percent,) yet we drive 35 percent of the world's cars and trucks. We like to drive alone in our convenient, prestigious cars, with our own music and our preferred level of air conditioning, rather than to ride together on a bus or a train. If we add up all the miles that Americans drive in a year, the total is more than all the miles driven by all other nationalities in the world combined. World oil consumption at present averages 77 million barrels a day, and the lion's share of that oil is burned in America.

We've taken the tobacco industry to court because of its poisonous smoke. Perhaps one day, the oil and automotive industries will be brought to a similar reckoning.

Not everyone wants to live in the Dark Ages. A growing number of European cities, especially in Italy, have car-free Sundays: no cars are allowed in the downtown areas. In a recent poll, 80 percent of the French respondents said they would like car-free days on a regular basis. People enjoy the quiet urban parks, the cleaner air, and the opportunity to talk with each other on the street, while shops and cafes profit from an increase in business.

Many cities around the world encourage the use of bicycles. They build bike lanes not only within the city, but parallel to highways running from city to city. In Copenhagen, people can borrow bike much the same as they borrow a luggage trolley at the airport: they put a coin into a box on the rack which holds the bike, then ride off to wherever they want to go. When they return the bike to the rack, (or to any comparable rack,) they receive a coin back. Scandinavian countries, as well as the Netherlands, Germany, and Japan, enable people to take their bikes with them on trains and buses, so that the bike can be combined with public transportation.

Many countries around the world have invested in mass transit, which is a far more efficient way to move people and freight, with regard to both economics and ecology, than individual cars. By boat, bus, and train, a person can easily travel on public transportation from a remote fishing village in northern Norway, far above the polar circle, to the international airport outside of Oslo. On the other hand, a person trying to travel by public transportation from Lake Placid to Kennedy Airport will confront an exasperating lack of connections, as if buses, trains and planes had never heard of each other. With mass transit, children, the elderly, and those who cannot afford a car can still get where they want to go. But in this land where the Car is King, parents must drive their kids everywhere, the elderly are often stuck at home, and poor folks save their money to buy a rust bucket.

The multiple costs from burning oil are substantial on today's economic and ecological tally sheets. But our reliance on oil has a future cost as well.

Many other countries are investing in the research and development of alternative fuels, alternative vehicles, alternative power generators, at a far more serious level than that found in the United States. Japan is now mass producing hybrid cars that run on both gasoline and electricity: the gasoline engine recharges the battery while the car is on the road, and when the battery is ready, it takes over. The engine thus switches back and forth. Methane and ethanol will eventually replace gasoline in the hybrid engine, making it even cleaner.

Likewise, other countries are investing far more into hydrogen as a future fuel, made from solar power and water. Saudi Arabia, the world's foremost oil producer, is investing heavily in hydrogen research. The sun has become a business partner in dozens of countries, where the oil and automotive industries do not have such powerful lobbies knocking on the doors of government. Or making fat, quiet campaign contributions.

The time will come, hopefully soon, when clean cars and clean power generators will replace the dirty machines we now rely on. The innovative companies which manufacture those cars and generators will sell them to the rest of the world. They will hold the patents, and they will employ the researchers and engineers who are proud to be pioneers advancing into the future of global energy. Meanwhile, we Americans, with our minimal investment into renewable energy, can turn the empty factories and surrounding slums of Detroit into a Museum of Missed Opportunity.

Perhaps our American attitude is best portrayed by the multitude of car ads which show a Jeep or a SUV or a pick-up perched on top of a mountain peak, or parked beside a wilderness lake. The Jeep promises freedom, escape, manhood, the rough-and-ready lifestyle.

Never mind how that bright red Jeep got to the mountain top, what it did with its four-wheel traction as it roared up the slope, or what the other thousands of Jeeps which it is promoting are likely to do to that mountain. Never mind that the snow on the peaks in the background is probably acidic. Never mind that the water in the lake is most likely acidic as well, and that the trout photographed in the frying pan over the campfire were brought in a cooler from a supermarket down in Denver. Never mind that the advertising agency photographer waited three days for the air to clear before he took his picture of the all-conquering Jeep.

We have mastered the mountain with our Jeep. No wilderness lake is too remote for our SUV. Our snowmobiles can cross a snowy meadow at eighty miles an hour.

The natural world is our playground.

We're the kings of the Dark Ages.

But the revolution, the new way of thinking about cost, and health, and democracy, is quietly coming.

Funeral Veil

Were Earth an increment closer to the sun,
No snow would there be, nor me.
Were Earth an increment further from the Golden One,
No rain would there be, nor me.

Wherefore then, in perfect and perpetual equilibrium,
Do we earthlings yearly conceal ourselves behind
A shroud ever darker,
Blocking with all manner of airborne muck
The gentle rays that would bless us, bless us, bless us unto eternity?

Shall we not allow the sun himself to bestow his spark
Unto our multitude of clamoring machines,
Before the Earth herself is forced to wear the veil
Of a mother mourning for her children?

VILLAGE OF
REMSEN
POLARIS
POLARIS
TIGERSHARK
VENTURE

Invitation to a Renaissance

Let's let a little fresh air blow through this disconcerting book. A little wind.

Let's let a little sunshine banish the shadows of doubt and despair.

Electrical current is made when a magnet spins inside a coil of copper wire. There are many ways to make the magnet spin: attached to an axle, which in turn is attached to wheel with blades, the magnet spins when the blades are turned by a flow of water, a flow of steam, or a flow of wind.

Hydroelectric dams back up a river, then channel the flow of water so that it turns the blades which drive the electrical turbines. A flow of water thus engenders a flow of electricity over wires that spread across the land. But there are only so many rivers that we can dam, or would want to dam, for disrupting the flow of water, and especially backing the water into a large lake, can have major adverse consequences. Upstream, farmland, residential land, and wilderness land may be flooded; downstream, fluctuations in the flow of water can threaten populations of both people and wildlife.

In a coal- or oil-fueled power plant, the burning fuel heats water into steam. The steam drives the blades, which turn the magnet inside the turbine. But the burning of coal and oil, as we have seen, releases enormous amounts of sulfur, nitrogen, and especially carbon into the air.

In a nuclear-powered plant, radiation heats the water to steam, which then turns the turbines. However, though most nuclear power plants have proven to be safe, we still have not agreed upon a safe repository for increasing amounts of lethal nuclear waste. No one can guarantee that nuclear waste buried deep underground in a desert won't be disturbed a century from now by an earthquake, releasing radioactive waste into the earth, and possibly into the ground water and the air. With so many nations creating more and more nuclear waste, which is kept in a variety of places around the world, how can we be sure that all of that waste, everywhere, is safe? For if any of it leaks, it becomes everyone's problem.

Even the most sophisticated coal-, oil-, and nuclear-powered plants are doing nothing more then heating water to make steam. We've upgraded the fuel, but we're still using James Watt's steam engine of 1765 to turn the axle.

We can certainly run out of rivers to dam, but we'll never run out of wind. And wind releases not a whiff of smoke as it turns the axle. Nor do we need great quantities of wind to produce sufficient usable electricity: recent research has shown that enough Class 4 wind (wind of a certain strength) blows across the plains of North Dakota to produce 36 percent of the electricity required in 1990 by the forty-eight contiguous states. If we add the winds of South Dakota and Texas, we could produce more than one and a half times the electricity used by the entire country in 1990. And every kilowatt is clean.

California produces the most wind-powered electricity in America, followed by the states of Minnesota and Iowa. Farmers in Iowa are now earning an average of $2000 per year for every wind turbine they erect on their land. Each wind generator takes up so little room, only a few square meters at the base of the tower, that the farmers can continue to plant their crops, and cattle can continue to graze.

European countries are investing far more into wind power than is the United States, and consequently they are developing new technologies at a more rapid pace. Denmark is the world's foremost producer of wind turbines. The coastline of Sweden is sprinkled with spinning white windmills; the large Swedish island of Gotland produces almost all of its electrical power from wind generators. Germany and Spain are rapidly increasing their wind capacities.

In 1999, wind power was the world's fastest growing source of electricity. Wind turbines across India are bringing electricity to remote villages, and other countries around the globe are providing electricity to their rural villages simply by erecting a single wind generator in each village.

As the market expands, those countries with the technology will be selling their fiberglass blades and quiet turbines to global customers who need a single generator or an entire wind farm.

Wind power can be coupled with electricity generated by traditional fuels, so that whenever the wind blows, the electricity bill drops significantly. Some homes which are wind powered are actually able to sell their excess power to the local utility; as their kilowatt wheel spins backwards, the power company pays them the going rate.

Like the wind, the sun had long been a willing friend, but a friend that we have all but ignored.

In 1979, President Jimmy Carter installed a solar water heater on the roof of the White House, in part to heat water, and in part to announce to the nation that America was embarking on a new era of energy use: finally, we could accept the gift offered to us daily by the sun. The Carter Administration initiated tax credits for new homes built to take advantage of heat and light from the sun. Solar research advanced at a steady pace, and small businesses offering solar products were growing like young saplings.

That four-year burst of research and product development, a period during which the American pioneering spirit was reborn, was followed by eight years of cuts in research funding, and an end to tax credits for innovative housing. President Reagan took down the solar water heater from the roof of the White House, and taxpayers once again paid for oil to heat the President's bath water.

Other countries are advancing steadily in developing a beneficial relationship with the sun. At the same time that Reagan was cutting funds, Japan, Germany, and Italy were investing in solar research. All three are free from pressure by domestic oil companies, and thus their first priority is not the welfare of the stockholders.

As they are doing with wind turbines, rural villages around the world are presently installing systems which collect sunlight and turn it into electricity. Villagers can work and study during the evenings. Craftsmen can earn a living where they live; they no longer need to migrate to the slums of some huge city to seek work. The market for this solar technology is enormous, much greater, for example than the market for laptops.

Like wind-powered generators, solar panels produce electricity without any smoke laden with sulfur, nitrogen, and carbon. Yes, in the initial manufacturing process of windmills and solar panels, a limited amount of pollution is produced. But it is far less than the daily outpouring from smokestacks and spent nuclear rods.

America's future, and the Earth's future, are threatened by the tenacious grip of the coal industry, the oil industry, the nuclear industry, and the automotive industry on the way we produce electricity, and the way we travel. America may build the fastest computers and the smartest weapons, but the real winners during the next century will be those nations that team up with the sun.

At the beginning of the 1960's, President John Kennedy challenged the nation to put a man on the moon by the end of the decade. America's spirit was kindled; we watched as NASA reached further and further toward our goal, until in the summer of 1969, the Eagle which was launched from Florida made a precision landing on the moon. Neil Armstrong left his historic footprints in the moon dust, and Americans felt that we were well on our way toward . . . toward what? Something which very few of us could really participate in.

How much better for a President today to challenge the nation to produce half of its electricity from the sun and wind by the end of the decade. How much better to fund universities rather then rocket ships, to fund a multitude of small businesses rather than a few aerospace giants, to fund a truly national pioneering effort, rather than a cluster of experts in Houston. How much better to replace tons of carbon blowing on the wind, with a clean wind that produces electricity. And certainly, how much better to give our kids a real challenge, a national challenge, a global challenge, rather than to park them in front of the latest sit-com, the latest action video, the latest computer game. Fifty-seven percent of America's electricity comes from coal, and 60 percent of American kids have a television in their bedroom. Let's change that. Yes, the kids can tinker on the computer that they love, but let them tinker with a real purpose: to make the Earth the healthy place it once was, to provide the blessings of electricity to one and all, and to prosper economically while doing so.

And while we're at it, let's set another national goal: by 2010, let's be well on our way toward making transportation in America clean. If we can put a rocket on the moon, we can build a clean car. A clean bus. A clean truck. A clean public transportation system. President Johnson put the headquarters of the space program in Texas. President Reagan put the bulk of military production in California. Let's put research and production in public transportation in every state. Let's turn the inner cities into hives of progress in transportation.

If the Oil Boys start to weep again, we can loan them a hankie.

We need a President, an administration, and a Congress which can do more than bicker, pitter-putter, and make buzz-word speeches. We need a President with stature enough to turn away from the Oil Boys, so that he can lift his face toward the sun, and take a good deep breath of the wind. We need a President who has the courage and the humility and the wisdom to ask: What are the electrical engineers doing in Denmark? What are they doing in India? We need a President who understands that coal and oil came from plants millions of years ago, plants which spread their leaves to the sun.

If I may say so, we need a President like Harry Truman, who had more honesty, integrity and guts then anyone since.

Harry Truman had the gift of liking ordinary Americans. He truly enjoyed their company, and he thought about the welfare of the farmers and the shopkeepers. At the same time, he had the gift of reaching beyond Normalcy, toward a time when life could be better for all kinds of people, including people living in other countries.

Yes, he knew about planting corn and raising hogs, he loved to drive a car, he loved to ride on a train, he loved to be a snappy dresser.

But he also knew the history of France when he arrived there as an artillery officer during World War I, because, mostly on his own, he read books.

As a senator from Missouri, he led a committee that shone its searchlight at corruption on a national scale, in both pricing and production, by American war profiteers during World War II. Despite their flag waving, a number of corporate executives were not as patriotic as they would like to appear, even to the point of producing defective airplane wings which they knew to be defective, in order to turn a profit. Until Harry nailed them.

As President, Harry Truman was the main force behind the Truman Doctrine, which was an attempt, often successful, to protect the countries of Europe from Stalin. His administration initiated and pushed through Congress the Marshall Plan, which enabled nineteen European countries, in ruins after World War II, to get back on their feet. Harry Truman helped to create NATO, America's first long-term military alliance, with the hope of avoiding World War III. Truman directed the Berlin airlift, which from June of 1948 to September of 1949, fed and clothed and heated the people of Berlin by sending in more planes than had ever been landed in one place in history, . . . until Stalin backed down and removed his blockade of the city.

In his inaugural address in 1948, Truman explained his Four Points of foreign aid, which promised the developing countries of the world not simply money, but education, shared expertise, technical training, and above all, help with building new schools. He was the first President in American history to ask Congress for civil rights legislation, based on the recommendations of a committee which he himself set up. He recognized the new nation of Israel, against fierce opposition from people who did not want to anger the Arabs, because the Arabs had oil.

The list goes on, but the point is: President Harry Truman was always searching for the new idea, the new and better way of doing things, for everyone. America was prosperous after World War II, and he could easily have sat at his presidential desk and cheered, "Ain't we got it good!" Instead, he reached out and tried to help people, people all around the world.

Because he helped other nations to get back on their feet economically, and to become better educated, those nations have helped America's international trade to grow ever since. And those nations have helped to maintain world peace.

Of course, a President cannot do it alone. He or she can't do it with a population that spends big bucks for loophole vehicles. He can't do it with a population willing to send the troops now and then to the Middle East, there to lose a few good men and women so that we can enjoy cheap gasoline at the pumps back home. He can't do it if Congress maintains its stranglehold on funding for research in alternative energies. He can't do it if we Americans continue to think that the threat of global warming is just so much hot air. No American President can steer the nation toward ecological and economic prosperity until the voters, that great amorphous group of people over eighteen years old, truly begin to care about the kids, that great amorphous group of people under eighteen years old.

Because it ain't just about how much we pay per kilowatt on the monthly bill. It's also about incarcerating our kids in a continuing Dark Ages, or offering them a genuine Renaissance.

Circle

On Earth was born a circle.
Absent most elsewhere in this fired and cindered universe,
The initial circle here was an intangible force that gathered,
And grew stronger as it gathered,
Until gravity revealed her secret: the world itself,
A rough experiment in circulature
Until drops fell into a pond.

Daisy, spinning in your fingers.
Trillium, pointing to the existence of an equality within thirds.
Iris, a confabulation of curves.
So the circle developed,
Until iris again, it encircles our stare.

Pioneers, or Pirates?

As a teacher of twenty-five years, I care deeply about kids.

So this is a book about more than just acid rain.

This is a book about our children. Not my children, not your children, but *our* children. Do we educate them to be oblivious, or to be mildly concerned; or do we educate them so that they can gladly tackle the challenge of curing the sickness in the air and water of the land they live in?

Do we educate them to think of my Adirondacks, my Vermont, my Virginia, my Ohio, my Illinois, or do we awaken them to the fact, and the responsibility, that we all live in the same nation?

Do we educate our children to ignore other children in other lands, or do we teach our children that they, like Canadian children, Norwegian children, Swedish children, Finnish children, Russian children, British and Dutch and German and Greek and Chinese and Japanese and Argentinian children, all of whom are threatened today by the effects of acid rain, are fellow citizens of the same world?

And of course, do we bestow upon our children, and their children, a healthy inheritance, or do we wish them well as they struggle to survive in a graveyard?

Let us question the nature of our patriotism. Because whether the problem is acid precipitation, or strip mining of coal, or clearcut timbering and the subsequent erosion, or mismanagement of radioactive waste, the fundamental question remains the same: do we love and honor and protect the American land as a sacred gift, a gift shared by all who live here, saplings and fingerlings as well as children; or do we desecrate our forest, our nation, our world, like barbarians plundering a temple?

Do we continue to harness our American economic system to scattered and limited energy sources, often in unstable regions of the world, or do we harmonize our technology with the sun, the wind, the tides, the geothermal heat in the Earth, upon which no one can claim a patent or ownership?

Do we continue to teach our children that money today is better than health tomorrow? That a dead lake is still good for water skiing? That hamburgers (from cattle raised where the Brazilian rainforest was cut down to make room for grazing) taste better than perch or bass or trout? Or do we teach our children that by the end of this new century, they must balance the right number of people with the right number of trees, or they may well have neither of both?

Do we continue to force more and more of our children to contend with asthma, a rapidly growing public problem, caused by sulfur dioxide particles in the air from the burning of coal?

Do we really care in the United States of America more about our economy than about our kids?

Do we rise to the occasion, do we muster the courage, do we embrace the challenge and exhilaration of a true renaissance; or do we continue to stumble and bumble along, squabbling in Congress, bickering the United Nations, content that we got through one more summer with Fourth of July fireworks and hotdogs on the grill?

Do we honor the Creator by honoring His Creation, or do we put a five dollar bill into the basket on Sunday morning, then pay with a twenty dollar bill at the gas pump?

Do we live up to our calling as We the People, or do we continue as We the Consumers, We the Shareholders, We the Corporate Board Members, We the Superpower?

Is there a spark of divine will in the heart of every person, or are we just a bunch of lucky brats?

Do we lift our faces to the sun, which over the centuries has sustained us every step of the way, or do our faces grow pale and our eyes bleary in front of a television screen, a computer screen, a video screen?

Are we pioneers, or are we pirates?

Do we teach our children to explore, or just to ask for more?

Acid rain, acid snow, and acid clouds have so altered the chemistry of the air, land, and water of the American Northeast that we are now able to observe firsthand the process of extinction. The extinction not just of a species, but of an entire ecosystem. Because once that rich, intricate, delicate, beautiful, irreplaceable forest growing on a mountain peak is dead, as many now are, the soil that lies upon the rocks washes downhill with every rainstorm, every snowmelt, until little is left but rock. Then people can hike up the 46 high peaks of the Adirondacks so that they can take pictures of 46 tombstones.

If the chemistry in your blood were as altered as the chemistry of the streams that flow through the American Northeast every spring, you would be dead.

Let's tap the energy and the optimism and the talents of our kids. Let's show them that we believe in them. Let's replace their restlessness and frustration with a sense of purpose.

We can give them no finer gift as we enter the new century.

Star Dust

The slums of the world are one slum.
From the same dust of the same exploded star,
We spun and slowly gathered into a small ball
Orbiting around a larger hot one.
The dust beneath our feet
Is the same as the feet themselves.
This bread we break or do not break with one another:
Star dust.
This wine in fellowship or sanctity or loneliness partaken,
Once reached in radiant expanding waves of dust across the universe,
The heavens,
The dark.
Who are we not to share?

A Global Prescription

So what do we do? Where do we start?

First, let's do away with the myth that cleaning up our mess is too expensive. As noted earlier, the people who run coal- and oil-fueled power plants protested, in opposing the 1990 Clean Air Act amendments, that the costs of compliance with the new standards would be enormously high. Some utilities hired public relations firms, which sent out mailings to electricity consumers stating that their bills would increase by as much as 30 percent if the Clean Air Act amendments were passed by Congress into law. In fact, the actual costs have proven to be less than one-tenth of the Big Coal and Big Oil estimates. How much longer do we listen to the cry babies in Gucci shoes?

Further, health benefits to the American people have far outweighed the costs. People who breathe dirty air and drink dirty water become sick, stay home from school and work, and run up doctors' bills; people who breathe clean air and drink clean water stay healthy, and miss few days of school and work. In 1996, the Environmental Protection Agency released a study of the costs and benefits of clean air regulations between 1970 and 1990. During that twenty-year period, (the first two decades of a serious effort in America to provide clean air for its people,) the total cost, including the cost to electrical power utilities, was roughly $436 billion. However, the benefits to human health were calculated as well above three trillion dollars, and possibly as high as fourteen trillion. Clearly, a healthy economy depends on healthy people.

European figures parallel American figures. In a broad range of countries, health benefits have significantly exceeded the costs of pollution control.

The economic analysis can be taken a step further. As we have seen during the computer revolution, new technologies create new jobs, new businesses, and increasing opportunities in research and development. Computer products created and built in the United States can be shipped abroad. So it could be with wind and solar products. Digging coal in West Virginia and burning it in Ohio is not providing an abundance of new jobs for American workers. But were the American government to provide the same amount of financial support to the pioneers of renewable energy that it has long provided to the fossil fuel corporate giants, and to the nuclear giants, then renewable energy would offer exciting and meaningful careers to an ever-growing number of people. It's time to put Goliath on a diet, and to offer David a helping hand.

Second, it's time for *us* to accept true responsibility for cleaning up our mess. Recycling newspapers and aluminum cans is not enough. Sports Utility Vehicles and other lightweight trucks take advantage of a legal loophole, a loophole created and defended by the automotive industry, which allows that type of passenger vehicle to get fewer miles per gallon, and thus to produce more pollution per mile. The SUV is dirtier than the standard sedan; we all know it, and yet nearly half of the cars bought today are lightweight trucks. When you're ready for your next car, tell your friendly neighborhood dealer to send his FLV (Filthy Loophole Vehicle) back to Detroit. You'll take a car that quietly speaks of your integrity, please.

Do we really need to teach our kids that they can't have fun unless they're burning gasoline? Trails in the Adirondacks are increasingly torn up by kids riding along the forest paths on their dirt bikes and four-wheelers. A dozen canoes on a lake do not disturb a single person's Adirondack afternoon, but one roaring jet-ski, zigzagging back and forth like a giant insect with wings and no brain, rudely ruins the afternoon for nearly everyone on the lake. Our kids are learning not only to ignore the fumes, but to ignore their neighbors. We are teaching a mindless reliance on oil, and an oblivious, or even hostile, attitude toward other people.

In the old days, when a hunter returned with a deer to his village, he was respected to the degree that he shared his bounty with his neighbors. His manhood depended upon his ability and willingness to offer a meal to a widow, to children, to the elderly. Today we no longer bring a deer back to the village, but we might still offer a breath of cleaner air to our neighbors.

Our governments in Washington and in the state capitals have got to do their part, but they won't unless we nudge them, firmly. The present statute, with its limits and loopholes, will only prolong the dying of forests and lakes, rather then encourage their recuperation. The General Accounting Office, a research branch of Congress, stated in its report in March, 2000, that "Lakes in the Adirondack Mountains are taking longer to recover than lakes located elsewhere and are likely to recover less or not recover, without further reductions of acid deposition." The report concludes, "Increases in these lakes' acidity raise questions about their prospects for recovering under the current program." In other words, the sulfur and nitrogen limits set by the 1990 Clean Air Act amendments are not strict enough to stop the slow death of many more lakes in the Adirondacks.

In Washington, New York representatives Sherwood Boehlert and John E. Sweeney have each presented bills, H.R. 25 and H.R. 657 respectively, to the House of Representatives, asking for lower emissions limits for both sulfur and nitrogen, as well as regulation of mercury emissions, to be applied to all power plants.

New York Senator Daniel Patrick Moynihan has presented a bill, S. 172, with comparable goals, to the Senate.

Both of these bills are presently blocked by members of Congress from the Midwest, or from states which believe they have no acid rain problems.

Write a letter to your representatives in Congress, asking them to support H.R. 25 and 657. Write a letter to your senators, asking them to support S. 171 and S. 172. Those good people need to hear your voice, before theirs can be heard.

Write letters as well to other members of the Senate and House of Representatives who hold leading positions on environmental committees. You will find a number of important names at the end of this book.

People of the Adirondacks, people of Appalachia, rise up and roar.

Third, we need to join the international effort. When I visited the Norwegian Institute for Air Research near Oslo in January 2000, I listened to three researchers speaking at length about their multi-country network of sampling stations and labs, a network reaching throughout Europe and into Russia. Two hundred fifty-six cities in 22 countries, with a total of 896 sampling stations, now work together in a system called "Euroairnet."

When I read a Swedish quarterly journal called "Acid News," I find articles about projects and progress around the world: the effect of acid rain on the Taj Mahal, emissions standards for international shipping, the burning of peat in Estonia, electric vehicles in France, a car-free day in Cracow, zero-emissions vehicles in California, bicycles in the Netherlands, traffic patterns in Japan, nuclear power in Bulgaria, wind power in Spain, solar power on Crete, nuclear waste on Russia's Kola Peninsula, ozone in Indonesia. The articles are concise and clearly written; in half an hour, I have an update about pollution problems and progress around the world.

On the other hand, during my interviews in America, research scientists and political officials spoke almost exclusively about the situation in the United States, with a mention now and then of Canada. American publications focus on American research, American legislation, American economics, with a brief mention now and then about acidified lakes in Norway and Sweden. Canadian articles are clear and readable; American articles tend to be highly scientific, clogged with statistics, symbols and jargon.

I am left with the feeling that Europeans and Canadians know there is a world out there, whereas Americans think that their continental island is *IT*.

We need a traveling Ecological Exposition, an international trade show with exhibits of the latest in environmental technology from nations around the world. The show would move to twelve locations around the globe, staying at each for three weeks, (and taking a week to move to the next location,) so that during the course of a year, people of every continent would have the opportunity to see what other countries are doing to make the world, our world, a healthier place for our children.

Let's say that the Ecological Expo sets up in an exhibition hall in West Virginia, not far from a coal strip mining operation. During those three weeks, busloads of kids from all around the eastern half of the United States could attend; university faculties could attend; small business entrepreneurs could attend; governors and mayors could attend; church groups could attend; and we would all discover the startling progress that has been made in Denmark and India in the field of wind turbines, the progress in solar hot water heaters in Israel, the progress in solar-hydrogen fuel in Saudi Arabia. When our feet are tired from walking from booth to booth, we can get on a tour bus and ride out to the strip mine, where we watch giant machines shoveling up acres of coal. Maybe somebody at the site will sell souvenir boxes of coal soot.

Fourth, every school needs more than a chapter on ecology, followed by a ten-question quiz. The kids need to put on their boots and take their classroom pH meters out to the lakes and into the forests.

Here in the New York Adirondack Mountains, we have hundreds of lakes and ponds, hundreds of brooks and streams and rivers, and countless bogs and springtime pools, but only a few state researchers with very limited funds for testing this vast and intricate web of water. How easy, how inexpensive, how productive, and how meaningful it would be if the high schools in the Adirondacks, in Appalachia, in the Rockies, in the Los Angeles basin, enabled their kids to contribute to a testing program. Let those kids pool their information on the Web statewide, nationwide, globally. No reason why an acid rain research project at a school in Lake Placid, or Long Lake, or Old Forge, can't cooperate with an acid rain project at a school in Murmansk, or Romania, or Brazil.

Kids have the energy, the motivation, the optimism. And from what I know of American kids, despite their overflowing toy box, they are starved for something to do that has a real purpose. What better purpose than to help clean up the world, the whole world, putting to an end the sorry mess which the old folks have made.

Fifth, you can give a ton of coal soot for Christmas. Current legislation in the United States enables coal-burning power plants to buy and sell sulfur dioxide pollution allowances. This system encourages utilities to produce electricity more cleanly, beneath their pollution limit, so that they can sell their unused quota to other utilities. The Niagara Mohawk Power Corporation of New York has donated a number of its allowances to the Adirondack Council, a conservation organization. For $50, you can purchase one ton of sulfur dioxide pollution, so that your ton will never go up a smokestack. You will receive a Clean Air Certificate, with your name penned on it in graceful calligraphy, and a one-year membership to the Adirondack Council, a group leading the fight against acid rain.

These tons of sulfur dioxide make great Christmas gifts. My 96-year-old Aunt Myrtle proudly displays her certificate on a cupboard door in her kitchen.

Kids at school can raise money to buy a ton of pollution. Graduating seniors could buy a ton of pollution as a gift to their school, to their community, to the world. If every graduating class of every high school in the Adirondacks and in Appalachia bought a ton, over the years, the tons would add up significantly.

Churches can buy a ton. Parents of kids with asthma can buy a ton. A Girl Scout Troop can buy a ton. Businesses can buy a ton. On behalf of the birds at your birdfeeder, those beloved chickadees and finches and warblers, you too could buy a ton. On behalf of the peepers, a ton.

What do I want for my birthday? Never mind a silk tie, forget the CD. I want a ton of crud.

You can contact the Adirondack Council by mail at P.O. Box D-2, Two Church Street, Elizabethtown, New York, 12932, or by phone at 1-800-842-7275.

The peepers and the peewees and the parameciums will thank you.

And so will your grandchildren.

People began to measure the world's temperature in 1860. During the next 140 years, the thirteen warmest years have occurred between 1979 and 1998. The four record-breaking years were 1990, 1995, 1997, and 1998. That's *now.* These clearly rising temperatures were not engendered by someone's political agenda; they were engendered by more than half a billion cars and trucks emptying their gas tanks into the air every day, and by plumes of coal smoke and oil smoke, and by forests burning to fields of ashes.

The Environmental Protection Agency reports that emissions of greenhouse gases in the United States have increased by 20 percent during the years 1990-96, so that the United States is now responsible for 23 percent of the carbon (CO_2) emissions in the world. Though 4.6 percent of the world's population, we produce nearly a quarter of its carbon dioxide, a major greenhouse gas. Why continue with coal and oil? If we could get to the moon in less than a decade, then surely we can power a light bulb right here on Earth with a bit of sunshine.

Researchers have driven long tubes into the ice of Greenland and Antarctica, in order to capture bubbles of air from previous geologic ages. The amount of carbon dioxide in our atmosphere now is much higher than it has been for the past 420,000 years. Glaciers are presently melting. Plant and bird species are moving north. Norwegian fishermen in their eighties talk of the days when winter was winter, with deep snow; now, a bit of snow, then it's washed away by rain.

Let's put all those people who still refuse to believe in global warming in a house with a fireplace that has no chimney. They can play poker at the kitchen table while the attic fills with smoke. They can play gin rummy while the upstairs bedrooms fill with smoke. They can go down to the basement and continue gambling while the first floor fills with smoke. When, and if, they come crawling out the door for a breath of fresh air, we'll talk with them again about the value of wind generators.

During this entire discussion, we have not mentioned the oceans, which cover nearly 71 percent of the Earth's surface. The oceans absorb enormous amounts of heat from the atmosphere, and from the sun, until they can absorb no more, and so release heat. The oceans also absorb great amounts of pollution flowing into them from rivers and estuaries, as well as from their coastlines. Too much heat, too much pollution, and the oceans are not well.

Algae in the oceans produces about 90 percent of the oxygen in our atmosphere. You can breathe, you can live, because tiny drifting plants in the ocean absorbed CO_2, and, in the sunlit process of photosynthesis, released oxygen, O_2. Forests produce oxygen as well, but the bulk of the O_2 which enables life to continue on this planet comes from oceanic algae. The first wail of a newborn child is a wail of algal oxygen.

If the oceans become too warm, that algae may die in increasing quantities.

Already some coral reefs are turning white, "bleaching," because the algae which lives symbiotically within the coral is dying. Coral reefs are underwater forests; many forms of life live in them. If we threaten the algae, we threaten hundreds of other species.

Including us.

The physician ceases her labors only when the patient is cured, . . . or when a cure proves impossible.

We, as Earth's doctors, have barely begun.

The Believer

Devils' Laughter they call the flame-orange tree.
Well, it be more;
Now listen, son, and look.
See that yellow maple? A ghost of shattered lightning.
Aye, and see the rich maroon? Your Mama' song at dusk.
That stand of aspen: shivering drops of a sleeping waterfall.
Mind for a moment that giant puffball, boy;
Each puff provoked by your probing toe
Sows a thousand potential progeny.
Then shuffle your boots through this cloak of tatters from the sun.

Those falling leaves be elves on magic carpets:
whirling
jigaliging
plunging (some do, you see)
pinwheel
scoot;
Winter-whittled, strewn like chippings of an October afternoon.

The boy spotted a gray cocoon on a twig.
Reaching up, he snapped off the twig
To keep the cocoon for next spring,
While his grandfather gazed without speaking
At the countable yellow leaves on the branches of a beech sapling,
Flickering faintly in the deep woods
Like the flames of old believers' candles in a sanctuary.

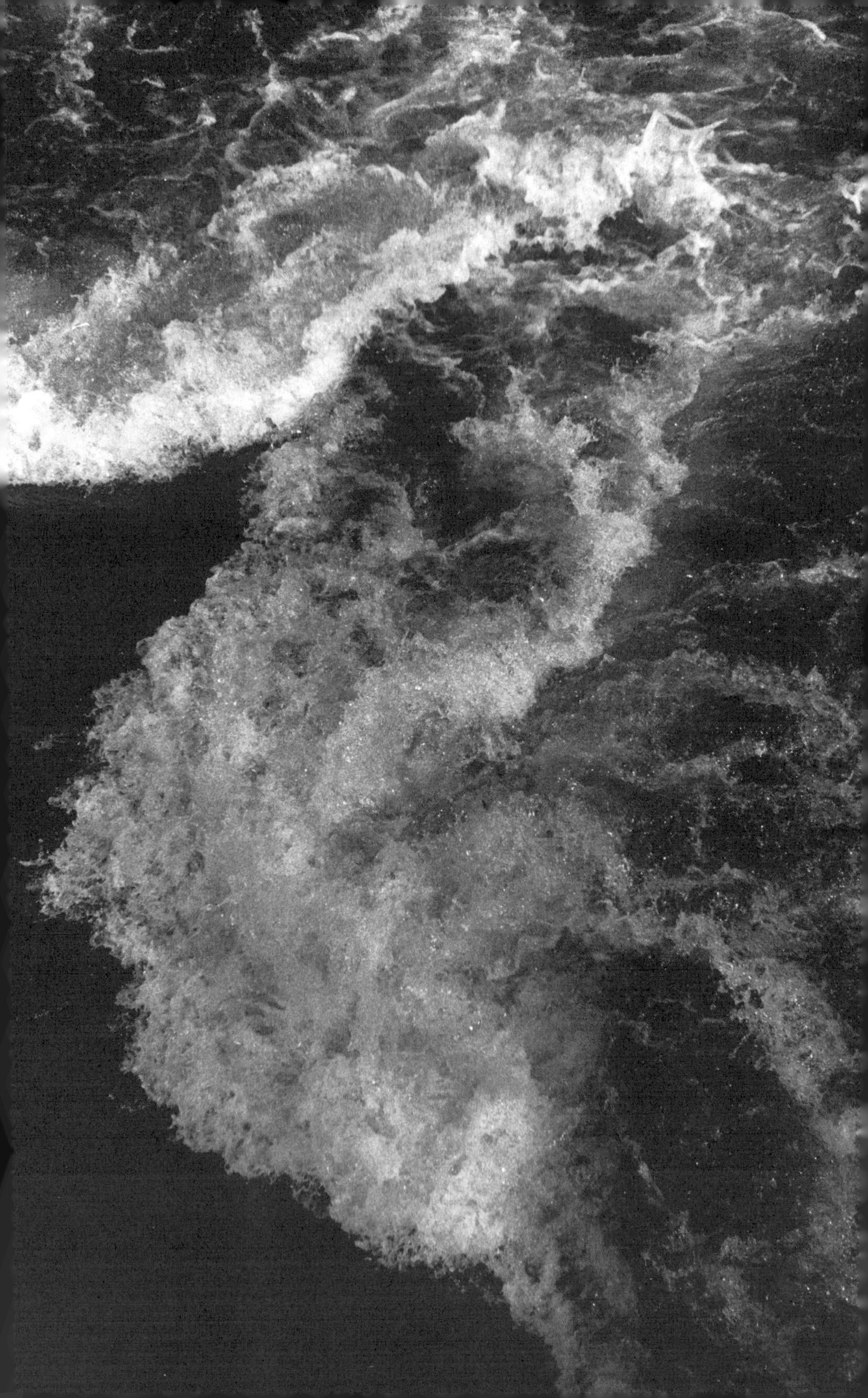

An Adirondack Symphony

I have often wished that I could write an Adirondack symphony. Johann Strauss the younger wrote his "Tales from the Vienna Woods," and Beethoven put his long walks through the Austrian countryside into his "Pastoral Symphony." Vivaldi poured his joyous heart into "The Four Seasons." Debussy takes us on a voyage in "La Mer." Ferde Grofé wrote his lyric and dramatic "Grand Canyon Suite," and Aaron Copland wrote his delicate, passionate "Appalachian Spring." Sibelius takes us deep into a northern forest in his "Finlandia." Tchaikovsky gallops us across the Russian steppes. Grieg even portrayed the Norwegian trolls who live in their mountain caves.

But no one to my knowledge has yet paddled with a notebook out to the middle of a sparkling Adirondack lake in August; and sat upon the peak of Algonquin Mountain as a huge black thunderstorm comes rumbling and flashing across a thrashing forest far below; and paused on skis in a snowy January woods to listen to big fluffy flakes sifting down through the dry brown leaves still clinging to the branches of a beech, . . . and then with these and a thousand other moments in mind, written a powerful Adirondack symphony.

I certainly can't do it. When I took piano lessons, my teacher asked me if perhaps I didn't have some talent in another direction, say, for example, a talent for playing the kazoo. I couldn't even write an Adirondack song, for I sing like a drowning turkey.

Perhaps I should hire a class of sixth graders for a day, take them out to a half dozen different places in the Adirondack wilderness, and ask them to write down as carefully as they can ten things that they hear. Then, while we eat our sandwiches on the sunny shore of a lake, each lad and lass would read a list of ten, (not just ten items, but ten sentences,) so that together we could begin to put what everyone has heard into some kind of order. Thus, in our symphonic poem, we would write the opening movement (we shall call it an overture) which portrays the wind whispering through a grove of tall white pines; and a breeze rattling through the leaves of a lakeside maple; and a rising gust washing tiny waves onto the lake's pebbly shore.

No doubt several students (boys, of course) will vocally contribute the roar of a chain saw, the incessant drone of a jet-ski, and their rendition of some favorite tune on a boom-box; but those troublesome lads we will send off on an expedition to record the sounds of peepers and grackles in a nearby wetland.

By now several girls are exploring a stream that gurgles down a rocky cascade. The stream becomes silent as it meanders into a slowly swirling black pool; it hums as it flows now past a fallen branch; and then (yes, it's true!) the stream disappears underground, or at least several inches deep, as it flows through glacially rounded Adirondack stones covered with a carpet of thick green moss, . . . the stream's gurgling is now muffled, mysterious, . . . until, fifty feet later, the stream reappears with its clear bright laughter.

A beautiful symphonic poem could that class of sixth graders write, and in doing so, they would have listened more carefully than they usually do, and they would have heard, in poetry and for real, what the other students heard.

The class will no doubt ask their teacher for a field trip up Algonquin Mountain, so that seated on separate boulders, each eleven-year-old with a notebook, the gang could prick their ears to the cry of a circling broad-winged hawk, to the chirp of a distant chickadee, and to the wind swimming through the crowns of trees far below.

Sounds of springtime, sounds of summer. The rustle of autumn; the deep silence of winter, which enables even a faint bluejay call to etch itself upon a blue-skied day. The symphony would grow and grow, with daily, weekly, monthly contributions from sleepy lads walking to the early-morning bus, and from clever lasses who hear the first flock of geese heading south, heading north.

Such a symphony, with a score of twenty parts, interwoven, leading from theme to theme, to be read aloud by twenty students, twenty authors, twenty poets, under the leadership of Maestro Teacher, might bring us all a long way toward understanding Maestro Nature's gentle, passionate, and very, very delicate choir of earthly voices.

Genesis, and Revelation

In the beginning God created the heaven and the earth.
And the earth was without form, and void;
and darkness was upon the face of the deep.
And the Spirit of God moved upon the face of the waters.
And God said, Let there be light: and there was light.
And God saw the light, that it was good:
and God divided the light from the darkness.
And God called the light Day, and the darkness he called Night.
And the evening and the morning were the first day.

Toward the end, the people covered the heaven and the earth with soot.
And the people had for so long turned their backs to the light,
facing with their pale countenances and eager hands
toward the coal and the oil
and the trees of the forest,
which they cut to the ground in great swathes across the land,
that the heaven became darkened with smoke,
and the light became like unto darkness,
until the earth, enshrouded in smoke, became increasingly void.
And the people called the fading light Twilight,
and the darkness they called Night,
And the Twilight and the Night were the first day.

And God said, Let there be a firmament in the midst of the waters,
and let it divide the waters from the waters.
And God made the firmament,
and divided the waters which were under the firmament
from the waters which were above the firmament: and it was so.
And God called the firmament Heaven.
And the evening and the morning were the second day.

And the people said, Let the firm be our temple,
let the corporation be our cathedral,
and let the money pour down like a mighty stream: and it was so.
And the people called the firm FirstStar Max.
And the twilight and the night were the second day.

And God said, Let the waters under the heaven
be gathered together unto one place,
and let the dry land appear: and it was so.
And God called the dry land Earth;
and the gathering together of the waters called he Seas:
and God saw that it was good.
And God said, Let the earth bring forth vegetation,
the herb yielding seed,
and the fruit tree yielding fruit after its kind, whose seed is in itself,
upon the earth: and it was so.
And the earth brought forth vegetation,
and herb yielding seed after its kind,
and the tree yielding fruit, whose seed was in itself, after its kind:
and God saw that it was good.
And the evening and the morning were the third day.

And the people said, Let us drain the wetlands, let us fill the swamps,
let us divert the rivers, even unto a trickle in the desert;
Yea, let us buttress the coastline with concrete: and it was so.
And the people called the dry land Property, Real Estate, and Acreage;
and the unwanted waters they called Run-off, and Sewage,
and Storm Surge;
and the people saw that it was profitable.
And the people said, Let the acreage bring forth industrial monocrops,
the herb yielding seed which will not sprout,
and the fruit tree yielding fruit doused with insecticide,
upon all the earth: and it was so.
And the earth brought forth brown fields beneath the weakening light,
and herb yielding seed that would not bear forth another generation,
and the tree yielding fruit that was poison like unto the bite of an asp:
and the people saw that it was very profitable.
And the Night and the Deep Night were the third day.

And God said, Let there be lights in the firmament of the heaven
to divide the day from the night;
and let them be for signs, and for seasons, and for days, and years;
And let them be for lights in the firmament of the heaven
to give light upon the earth: and it was so.
And god made two great lights;
the greater light to rule the day, and the lesser light to rule the night:
he made the stars also.
And God set them in the firmament of the heaven to give light
upon the earth,
And to rule over the day and over the night,
and to divide the light from the darkness:
and God saw that it was good.
And the evening and the morning were the fourth day.

And the people said, We need more electrical lights
in the growing darkness,
and more television screens and video screens and computer screens,
for as day becomes like unto night, so shall they comfort us,
and we shall watch our moving lights for days and for seasons
and for years,
and we shall see in them all manner of signs: and it was so.
And the people made two great lights;
the greater light, which they called Giant Screen, to rule the night,
and the lesser light, which they called Portamini, to rule the deep night:
they made the HandHelds also.
And the people set them in every room of every house
and waiting lounge and restaurant and classroom
to give light upon the sofa and chair and bed and desk
and park bench and airplane seat and bar stool,
that all these screens might rule over the night and over the deep night,
and divide the Almost Dead from the Dead:
and the people saw that it was a good write-off in the third quarter.
And the Deep Night and the Growing Cold were the fourth day.

And God said, Let the waters bring forth abundantly
the moving creatures that hath life,
and fowl that may fly above the earth in the open firmament of heaven.
And God created great sea monsters, and every living thing that moveth,
which the waters brought forth abundantly, after their kind,
and every winged fowl after its kind:
and God saw that it was good.
And God blessed them, saying,
Be fruitful, and multiply, and fill the waters in the seas,
and let fowl multiply in the earth.
And the evening and the morning were the fifth day.

And the people said, Let us pull enormous nets
behind our factory ships,
and let us pour our sludge into the bay,
and let us dynamite the coral reef, and anoint its fish
with cyanide,
that we may thereby take home a polyppy pet, and a bright darting clown;
and let us harpoon the great whales
that we may explode a bomb in their heads;
yea, shall the plutonium be submerged in rusty barrels abundantly,
until every living creature no longer moveth,
but each shall float and rot, or sink and rot, after its kind.
And every fowl with oil on its wings shall likewise succumb:
and the people saw that it was in their Contract of Dominion.
And the people addressed the codfish, saying,
Be fruitful, and multiply, and fill the waters in the filthy seas,
and let the fowl multiply in thousands upon thousands
of foul industrial coops,
that the corporate stockholders shall chant and call up, Hosannah!
And the Frigid Night and the Shrieks of Despair were the fifth day.

And God said, Let the earth bring forth the living creature
after its kind,
cattle, and creeping thing, and beast of the earth after its kind:
and it was so.
And God made the beast of the earth after its kind,
and cattle after their kind,
and everything that creepeth upon the earth after its kind:
and God saw that it was good.
And God said, Let us make man in our image, after our likeness;
and let them have dominion over the fish of the sea,
and over the fowl of the air, and over the cattle, and over all the earth,
and over every creeping thing that creepeth upon the earth.
So God created man in his own image, in the image of God created he
him;
male and female created he them.
And God blessed them, and God said unto them,
Be fruitful, and multiply, and fill the earth, and subdue it;
and have dominion over the fish of the sea, and over the fowl of the air,
and over every living thing that moveth upon the earth.
And God said, Behold, I have given you every herb bearing seed,
which is upon the face of all the earth,
and every tree, in which is the fruit of a tree yielding seed;
to you it shall be for food.
And to every beast of the earth, and to every fowl in the air,
and to every thing that creepeth upon the earth, wherein there is life,
I have given every green herb for food: and it was so.
And God saw every thing that he had made, and, behold,
it was very good.
And the evening and the morning were the sixth day.

And the people said, Let the living creatures upon the earth
make room for us,
and let the cattle eat hyssop and hormones,
that we may upon the flesh thereof
sprinkle the salt of the earth, and anoint with ketchup;
and let us exterminate every creeping thing,
that we alone shall creep, and crawl, and expire
in the Valley of Death which we ourselves have created;
and the people saw that nearly everyone was dead
before the living ran out of room to dig more graves:
and that was good.
And the people said, We had Dominion over Everything,
for in the image of God created we us,
male and female and every which way,
and we were fruitful and we multiplied until we filled the earth,
and we subdued every olive tree and mustard seed and lily of the field,
until, Behold, upon the face of all the earth
is there not one square cubit
left untrampled, unpoisoned, undug, uncut, unconquered
and unsquandered,
and the bones of all shall be as a testament to our sloth and our greed.
And shall a star shine brightly in the sky above us,
that other planets, other worlds, may follow it and so observe
in the course of their heavenly journeys
the sphere of soot and smoke, of silence and stench;
and look then upon themselves, still green and blue
and fresh and clean,
and blessed with the song of abundant life,
and know—these other planets, these other worlds—
that what in their wisdom
they have preserved and protected and praised in their prayers
is the best good of all good.
And the people saw not the Death upon the face of all the earth,
for they were of it.
And the Barren Cinder spun unto the sixth day.

Thus the heavens and the earth were finished,
and all the host of them.
And on the seventh day God ended his work which he had made;
and he rested on the seventh day from all his work which he had made.
And God blessed the seventh day, and sanctified it,
because that in it he had rested from all his work
which God created and made.

Thus the people were finished, all the host of them.
Yea, they ended their mighty work, and rested in the grave,
unto eternity.
And not a creature stirred which might have blessed, or sanctified,
or wept.

These are the generations of the heavens and of the earth
when they were created,
in the day that the Lord God made the earth and the heavens,
And every plant of the field before it was in the earth,
and every herb of the field before it grew;
for the Lord God had not caused it to rain upon the earth,
and there was not a man to till the ground.
But there went up a mist from the earth,
and watered the whole face of the ground.

These are the generations of the heavens and of the earth
when they were destroyed.
Now a rain fell upon the earth, a Black Rain,
over which stood no rainbow as a covenant,
and there was not a man to till the ground.
But there went up a pestilential stench from the earth,
the vapors of plague, the mist of burning earth,
and watered with the acid of sulfur the whole face of the ground.

And the Lord God formed man of the dust of the ground,
and breathed into his nostrils the breath of life;
and man became a living soul.
And the Lord god planted a garden eastward in Eden;
and there he put the man whom he had formed.
And out of the ground made the Lord God to grow every tree
that is pleasant to the sight, and good for food;
the tree of life also in the midst of the garden,
and the tree of knowledge of good and evil.
And a river went out of Eden to water the garden;
and from thence it was parted, and became four heads.
The name of the first is Pishon; . . .
And the name of the second river is Gihon; . . .
And the name of the third river is Hiddekel; . . .
And the fourth river is Euphrates.

And the Supreme People despoiled the dust of the earth,
and mankind in its folly did breathe into its nostrils the smoke of death,
knowing not that a soul hath no bottom line.
Their Eden withered, until naught stood above the darkened plain
but the flourishing tree of the knowledge of evil,
bearing fruit abundantly.
And a river went out of Eden to water the graveyard,
a river black with ash,
and from thence it was parted, and became four heads.
The name of the first is Nasdaq; . . .
And the name of the second river is Anthracite; . . .
And the name of the third river is Petroleum; . . .
And the fourth river is Axe.

A sample letter to your electricity company.

Your name and address

Your power company's name and address (found on electricity bill)

Dear (name of company),

Because of my interest in clean energy, I would be grateful if you would answer the following questions.

What percentage of your electricity is produced from the burning of coal? From the burning of oil? From nuclear radiation? From hydro-electric dams? From wind turbines? From solar panels? From biomass? From other sources?

Please explain why you use, or do not use, each of the above processes for the generation of electricity.

Do you sell pollution credits to other power companies? Are they located in our state of ___________, or outside of our state? Please explain your position on the sale of pollution credits.

What plans do you have for the production of electricity from renewable sources such as sunshine and wind? When will you begin actual production?

If you produce electricity from nuclear radiation, please explain what you do with the nuclear fuel when you are done with it. Does the nuclear waste remain in our state of _________, or is it transported elsewhere? How is it transported?

If you purchase electricity from any other power companies, electricity which you then distribute on your grid, please explain how those companies produce their electricity.

Thank you very much.

Cordially,

Your signature and
Printed name

Names and addresses of the good people in Washington, D.C. waiting to hear from you on the subjects of acid rain and clean energy.

In the United States Senate:

Senator Bob Smith (R-NH), Chairman
Environment and Public Works Committee
407 Hart Senate Office Building
Washington, DC 20510
(ph) 202-224-6176 - (fax) 202-224-5167

Senator Max Baucus (D-MT), Ranking Member
Environment and Public Works Committee
456 Dirksen Senate Office Building
Washington, DC 20510
(ph) 202-224-8832 - (fax) 202-224-1273

Senator James M. Inhofe (R-OK), Chairman
Clean Air Subcommittee
Environment and Public Works Committee
407 Hart Senate Office Building
Washington, DC 20510
(ph) 202-224-0146 - (fax) 202-228-2040

Senator Bob Graham (D-FL), Ranking Member
Clean Air Subcommittee
Environment and Public Works Committee
524 Hart Senate Office Building
Washington, DC 20510
(ph) 202-224-3041 - (fax) 202-224-2237

Senator Trent Lott (R-MS), Senate Majority Leader
487 Russell Senate Office Building
Washington, DC 20510
(ph) 202-224-6253 - (fax) 202-224-2262

Senator Tom Daschle (D-SD), Senate Minority Leader
509 Hart Senate Office Building
Washington, DC 20510
(ph) 202-224-2321 - (fax) 202-224-7895

In the United States House of Representatives:

Tom Bliley, Virginia, Chairman
2049 Rayburn House Office Bldg.
Washington, DC 20515
(202) 225-2815

Michael G. Oxley, Ohio
2233 Rayburn House Office Bldg.
Washington, DC 20515
(202) 225-2676

Michael Bilirakis, Florida
2369 Rayburn House Office Bldg.
Washington, DC 20515
(202) 225-5755

Joe Barton, Texas
2264 Rayburn House Office Bldg.
Washington, DC 20515
(202) 225-2002

Richard Burr, North Carolina
1513 Longworth House Office Bldg.
Washington, DC 20515
(202) 225-2071

Ed Bryant, Tennessee
408 Cannon House Office Bldg.
Washington, DC 20515
(202) 225-2811

John D. Dingell, Michigan
2328 Rayburn House Office Bldg.
Washington, DC 20515
(202) 225-4071

Rick Boucher, Virginia
2329 Rayburn House Office Bldg.
Washington, DC 20515
(202) 225-3861

Sherrod Brown, Ohio
201 Cannon House Office Bldg.
Washington, DC 20515
(202) 225-3401

Bart Gordon, Tennessee
2368 Rayburn House Office Bldg.
Washington, DC 20515
(202) 225-4231